baby CHIMPANZEES

MEG GREVE

CREATIVE EDUCATION • CREATIVE PAPERBACKS

CONT

ENTS

I AM AN INFANT.

I am a baby chimpanzee.

Look at my
long arms!
They help
me climb.

I weighed four pounds (2 kg)
when I was born.

I cling to my mom. I drink her milk. We sleep in a nest high in the trees.

I live in a group with
other chimps.
We wrestle and play.

We find food in the forest.

My face gets darker as I grow.

I walk when I am six months old. I sleep alone when I am five years old.

I will stay close to my mom until I am 10.

SPEAK AND LISTEN

Can you speak like a chimp?

Baby chimps scream and hoot.

Listen to these sounds:

https://www.youtube.com/watch?v=x7zkRHmA4oQ

CHIMP WORDS

chimps: a short way to say *chimpanzees*

cling: to hold on to

nest: a cozy spot for resting

wrestle: to pretend to fight

READING CORNER

Brett, Anna. *Little Chimpanzee: A Day in the Life of a Baby Chimp.* London: Words & Pictures, 2023.

Culliford, Amy. *Baby Chimpanzees.* Coral Springs, Fl.: Seahorse Publishing, 2022.

Garnett, Jaye. *Jane Goodall Chimpanzees.* Rolling Meadows, Ill.: Cottage Door Press, 2021.

INDEX

PUBLISHED BY CREATIVE EDUCATION AND CREATIVE PAPERBACKS
P.O. Box 227, Mankato, Minnesota 56002
Creative Education and Creative Paperbacks are imprints of The Creative Company
www.thecreativecompany.us

LIBRARY OF CONGRESS CATALOGING-IN-PUBLICATION DATA
Names: Greve, Meg, author.
Title: Baby chimpanzees / Meg Greve.
Description: Mankato, Minnesota : Creative Education and Creative Paperbacks, [2025] | Series: Starting out | Includes bibliographical references and index. | Audience: Ages 4-7 | Audience: Grades K-1 | Summary: "Introduce beginning readers to the treetop world of baby chimpanzees with this life science starter. Includes photos, a labeled animal diagram, "Make a Noise" section, glossary, and further resources"-- Provided by publisher.
Identifiers: LCCN 2024014682 (print) | LCCN 2024014683 (ebook) | ISBN 9781640264199 (library binding) | ISBN 9781628329520 (paperback) | ISBN 9781640005839 (ebook)
Subjects: LCSH: Chimpanzees--Juvenile literature. | Chimpanzees--Infancy--Juvenile literature.
Classification: LCC QL737.P94 G74 2025 (print) | LCC QL737.P94 (ebook) | DDC 599.88513/92--dc23/eng/20240419
LC record available at https://lccn.loc.gov/2024014682
LC ebook record available at https://lccn.loc.gov/2024014683

DESIGN AND PRODUCTION
Design by Rhea Magaro
Production by Beeline Media and Design, Inc.
Art direction by Tom Morgan

PHOTOGRAPHS by Alamy Stock Photo/dpa, 6, Fiona Rogers, 8, Martin Harvey, 4; Freep!k/gudkov, cover; Dreamstime/Patrick Rolands, 7; Shutterstock/AJancso, 14, Bohbeh, 13, Eric Isselee, 11, 12, Jane Rix, 9, Mario Plechaty Photograph, 2, Patrick Rolands, 5, 10

Printed in China